Brainse Dhromchonnrach
Drumcondra Branch
Tel: 8377206

# MOON *milk*

# MOON *milk*

## 55 PLANT-BASED RECIPES FOR A GOOD NIGHT'S SLEEP

GINA FONTANA

Hardie Grant
BOOKS

This edition published in 2020 by
Hardie Grant Books, an imprint of
Hardie Grant Publishing



Hardie Grant Books (Melbourne)
Building 1, 658 Church Street
Richmond, Victoria 3121

Hardie Grant Books (London)
5th & 6th Floors
52–54 Southwark Street
London SE1 1UN

hardiegrantbooks.com

A catalogue record for this book is available from the National Library of Australia

Moon Milk
ISBN 978 1 74379 593 4

10 9 8 7 6 5 4 3 2 1

QUAR.325433

Senior Commissioning Editor: Eszter Karpati
Senior Art Editor: Emma Clayton
Designer: Rachel Cross
Copy editor: Rachel Malig
Publisher: Samantha Warrington

# CONTENTS

# MEET GINA

I'm a certified health coach and health-food fanatic, and enjoy the creativity behind developing and photographing recipes that suit my lifestyle.

Having been diagnosed with celiac and thyroid disease a few years ago, I now follow a gluten-free, dairy-free, plant-based diet. By sharing my story and my recipes, I hope to help others like me, as well as those who simply seek healthy alternatives, through my blog. My drive to help people has fuelled me to work hard at perfecting my recipe development, food photography and writing. I want to show my followers that creating delicious, flavourful meals is possible, while still being an advocate for simplicity. My hope is to inspire people to pursue a healthier lifestyle, to become the best versions of themselves and enjoy all of life's moments without sacrificing indulgent foods, but rather finding a different way of making them. Moon milk has become one of my most searched-for recipes and many of my fellow bloggers have praised its health benefits, so I decided to put together a collection of clean, non-dairy moon milk recipes to allow anyone with a food intolerance to still enjoy this ancient drink.

I currently live in Columbus, Ohio, with my husband, son, daughter and two fur babies. Aside from being a busy mum running a small business, I enjoy playing softball, going to parks, singing, going to church and snuggling up on the couch watching my favourite evening shows, while of course enjoying a cup of moon milk. Thank you for allowing me to share these recipes with you.

I wish you many healthy blessings.

Gina Fontana

Gina Fontana
@HealthyLittleVittles

# AYURVEDIC PRACTICE

Ayurveda is an ancient practice that was founded as an Indian medical system thousands of years ago. It is based on a natural and holistic approach to physical and mental health, and treatment often combines mainly plant-based products in combination with exercise, massage, and specialized diets and lifestyles.

Sleep is a fundamental aspect of Ayurveda and considered just as important as diet in maintaining health and balance. Drinking warm milk as an aid to sleep is an ageless tradition believed to promote balanced emotions and ojas – the essential energy for body and mind that is gleaned from eating pure and nourishing foods.

The concept of drinking warm milk stems from the Ayurvedic belief that in order to digest milk properly, it should be brought to a boil, allowing it to foam up, thus changing the molecular structure so it's easier for human consumption. A pinch of pepper, cinnamon, turmeric or ginger was added to the warmed milk to reduce its heaviness and any mucus-causing effects. With the addition of a gorgeous array of superfood colours, the modern-day moon milk was born.

# SLEEP AND NUTRITION

It's no secret that sleep is imperative to our health - mentally, physically and emotionally. We all know that we could eat more healthily and exercise more, but sleep is also a fundamental piece of the puzzle.

Sleep is a necessary factor for healthy functioning, regulating our mood and our memory. Lack of sleep has also been linked to poorer diets and larger appetites, leading to health issues such as insomnia, sleep apnea, high blood pressure, high blood sugar and obesity.

So how does nutrition affect sleep? You may have a great bedtime routine in place - turning in at the same time each evening, and doing all sorts of rituals that you've read will help you achieve a deeper sleep, like keeping the room cool and dark, taking a shower before bed, putting away your phone and computer long before going to sleep or running an aromatherapy diffuser. But taking a closer look at our eating habits could be the key to sleeping optimally.

Many people grab a late-night snack, often loaded with empty calories, carbs and sugar, but what's actually happening is that they are waking their bodies up at the same time as trying to wind them down. This leads to overeating and packing on unwanted weight, while not supplying the body with any beneficial vitamins and minerals, and this plays a major role in sleeplessness. Having a calming moon milk before bed signals to the body that it's time to relax and prepare for sleep - and it tastes like a naturally sweetened, multi-beneficial, healthy dessert.

## MAINTAINING A HEALTHY DIET

While these moon milk recipes are intended to help you relax and get those ever-important zzzz's, as well as giving your body an added nutritional boost, it's also important to note the role of your daily diet. Including a wide variety of whole foods, fruits, vegetables, whole grains, quality protein (emphasizing plant-based sources) and good fats (olive oil, avocado, nuts and seeds) and limiting sugar intake and heavily processed foods plays a vital role not only in our health but in our sleep as well. Eating a variety of foods supplies an appropriate distribution of nutrients for sleep and gives the body what it needs to repair, restore and rejuvenate. Many of the recipes in this book include herbs, spices and superfoods that 'pick up the slack' if you've had an 'off' day, but they certainly don't replace or make up for other dietary choices.

# PLANT-BASED MILK

Plant-based 'milks' are certainly gaining in popularity these days. The dairy aisle at your local supermarket may now contain an abundance of milk choices, both dairy and non-dairy alike. It seems dairy intolerances are on the rise, maybe contributing to the growth of non-dairy milk sales and the decline of cow's milk sales. While it is no secret that cow's milk has a rich nutrient profile, plant-based milks can also carry a positive nutrient profile and be beneficial in our diet.

So how do you get milk from a nut, seed, grain or bean? Plant-based milks are made by grinding the nut, seed, grain or bean, then adding water, flavourings, vitamins and minerals. What's important to note when buying a plant-based milk as opposed to a dairy milk is that you should read the label to determine which brand is the best quality, containing a good nutrient profile and leaving out added sugars (unsweetened) and preservatives.

Plant-based milks include, but aren't limited to: almond milk - one of the most popular choices; soy milk - probably the most widely available plant-based milk; rice milk; cashew milk; coconut milk - becoming more popular due to its rich and creamy flavour; hemp milk; oat milk; flax milk; hazelnut milk; and pea milk - also gaining in popularity due to its high protein content.

You can also make your own plant-based milk at home with just a few simple tools. Soak your selected nuts in water overnight, then the next day drain, rinse and add them to a high-speed blender with water, vanilla extract (optional), salt and a natural sweetener (such as maple syrup - also optional). Blend until the nuts are completely ground up and the mixture is blended well. Then, over a bowl, pour the mixture into a nut milk bag or cheesecloth and squeeze the remaining liquid from the pulp. Store your milk in the refrigerator and use it in your favourite moon milk recipe. It's best to use the raw form of whatever you decide to make your milk out of - for example, raw almonds or raw cashews. You can also use the pulp in other recipes so it doesn't go to waste.

# MAKING YOUR MOON MILK

Moon milk is so easy to make, and each one follows a similar process. As you can see from the recipes in this book, moon milk is very versatile – the sky is the limit with flavour combinations.

## TOOLS

Perhaps one of the best things about making moon milk, other than its deliciousness and the amazing health and sleep benefits it possesses, is the minimal number of tools required to make it, meaning less fuss and more relaxation time sipping on these gorgeous evening 'lattes'.

You'll need a hob, a small or medium saucepan, a whisk, a measuring jug and spoons, and a mug (or two if you're generous enough to share) to whip yourself up any of these delicious warm milks. And if you like your moon milk frothy, you'll need a blender or a frother. That's it, really.

## HEATING

One of the main reasons why moon milk is considered such a powerful sleep remedy is because it is served warm. Moon milk is gaining in popularity these days for the comforting feelings it offers as you ease into your evening. Drinking a warm beverage before bed allows your body to relax more easily, getting you to dreamland quicker. Luckily, it takes very little time to heat up, so you can sit down after a long day and relax in no time at all.

Heating is an essential step to enjoying moon milk when intended as a pre-bedtime ritual. Typically, you heat your plant-based milk in a small saucepan over a medium-high heat. If you whisk it constantly while heating you'll obtain a frothier, more bubbly beverage, but you can also achieve the same texture by blending and frothing. Heating only takes about five minutes via the hob.

Heat your plant-based milk, whisk in your chosen ingredients, remove from the heat, froth if desired, then serve in your favourite mug and enjoy.

Note: the milk does froth up quickly as it gets hotter and can overflow from your saucepan, so watch it closely and whisk frequently. The frothing is a good indicator that your milk is ready to enjoy, but sip rather carefully as it will be very hot!

## ADDING YOUR INGREDIENTS

Moon milk is simple and doesn't require a lot of ingredients. You have your plant-based milk, then spice things up with milk-enhancing flavours: in some recipes, a superfood and/or an adaptogen - plants or herbs that are included in your diet to balance your hormones and adrenal system, helping the body cope with daily stress and insomnia - then either pure vanilla or almond extract, raw cacao butter or coconut oil and maybe a pinch of salt.

You'll notice that most of these plant-based moon milk recipes contain only five to six ingredients. Each one has been selected deliberately, with sleep and nutrition in mind, while also bringing something unique in terms of flavour, aroma and appearance. While some of these moon milks can be considered 'dessert', others are more savoury, with stimulating spices

## MAKING YOUR OWN FRUIT OR VEGETABLE POWDERS

Although it's easiest to buy your fruit or vegetable powders in shops or online, it is possible to make your own at home. There are a few different options.

Firstly, you can purchase the dried version of the fruit or vegetable, for example freeze-dried blueberries, then grind them in your food processor, coffee grinder or high-speed blender until a fine powder is achieved.

Another option is to buy the fruit or vegetable fresh, slice it thinly and dehydrate it using a dehydrator, then grind it up. If you don't have a dehydrator, place the fruit slices on cookie sheets and leave them in the oven overnight on the lowest setting.

The key is to make sure your fruit or vegetable has no moisture remaining. If you're not in a hurry and want to plan ahead, freezing your dehydrated produce overnight can also be very helpful.

You may choose to sift your homemade powder before adding it to your moon milk if it's a bit clumpy. You'll also need to experiment with the amount of produce needed to fulfil what the recipe calls for. It's recommended that you keep any leftovers in your refrigerator or freezer to avoid clumping and to keep any moisture away.

to help with a health ailment you might be experiencing. And feel free to experiment - these recipes can be followed exactly, or you can get creative with spice combinations and become a moon milk guru.

### BLENDING AND FROTHING

Achieving the beloved coffee-shop frothy latte experience is much easier than you think at home, especially when it comes to moon milk. These plant-based milks, when heated, become very frothy, especially when you whisk them continuously and rapidly. There's something about sipping a beverage that has been frothed, or in technical terms 'aerated', by beating air into the liquid, that gives an added touch of comfort.

But in case yours doesn't achieve a frothy consistency to your liking, you can also blend it, or use a frother after it's poured into your mug. A frother is a convenient, inexpensive handheld tool, so if you find you're really enjoying moon milk, it could be a great kitchen gadget to add to your collection. Your frother should come with instructions on how to achieve the best results, but generally moving the frother in a figure eight in your cup towards the surface of the liquid works best, though be careful not to fill your cup too much or it will spill up over the sides. The tiny bubbles formed during frothing make the milk increase in volume as the texture is made lighter and more airy.

## TOPPINGS AND DECORATING

Moon milk might not have been originally intended as the hip thing to sip; rather it was popular due to its functionality and medicinal capabilities in helping you calm your mind and move from night thinker to sound sleeper. But how can you not appreciate the gorgeous embellishments that make this one of the most chic beverages filling up your social media feeds? Decorating with toppings such as dried rose petals or flowers, sprinkled spices, freeze-dried fruits, nuts and seeds, drizzled maple syrup, a dollop of dairy-free whipped cream, sprinkles, coconut flakes or fresh herbs is both fun and relaxing, but of course optional. While you might enjoy adding these trimmings atop your warm beverage as a way to help you wind down after a long day (and impress all your friends online), the real treasure lies within your cup.

If you do decide to partake in a late-night art project, though, remember to keep the toppings proportional to the amount of moon milk in your mug. You don't want them to be overbearing. And if you'd like to enjoy your warm milk without intervals spent chewing the decorations, you may wish to omit them, although they can have their own added health benefits too. The choice is yours.

## ENJOYING FRESH AND STORING

While it's best to sip your moon milk right after you make it, you can also store it in the refrigerator and reheat it the next day. All of the recipes in this book yield enough for two people, depending on how big a mug you'd like to enjoy, so if it's moon milk for one, feel free to make the full recipe and store half in your fridge in a mug covered with cling film or in an airtight container for the next evening. You can reheat it in your microwave or in a saucepan, froth or blend, garnish and drink.

## SWEET AND SAVOURY

The following recipes have been developed and tested with taste, variety, visual appeal and, most importantly, sleep-aiding properties in mind. Some are sweet while some are more savoury. There are traditional dessert favourites, such as Carrot Cake and Coconut Cream Pie, which have been turned into healthy alternatives to the real thing and are designed to combat sleeplessness. Others, such as Ginger Turmeric and Chai, use ancient healing spices that will ignite your senses and help you relax. So let's dive right in.

## SERVING

Each moon milk recipe makes 500 ml (16 fl oz) of warmed milk - enough for two servings.

# ELDERBERRY ROSE

Well worthy of a double tap in your Instagram feed, this moon milk is one of the prettiest ways to unwind after a long, hard day, and it's one of the most immune-boosting too.

### Ingredients

500 ml (16 fl oz) unsweetened oat milk
1 tablespoon elderberry powder
1 tablespoon maple syrup
1 tablespoon dried rose petals (food-grade), plus extra to garnish (optional)

### Tip

If you can't find oat milk, you can use unsweetened almond milk in this recipe instead. And if you're not a fan of drinking the dried rose petals, simply let them infuse in the warm milk for an additional few minutes, then strain your milk before enjoying.

Elderberries have stood the test of time and are still one of the top antiviral foods today. They are loaded with antioxidants and anti-inflammatories, which is why you may have heard of taking elderberry syrup or supplements when you're sick as a natural remedy to help kick that cold.

Now you can add it to warm milk infused with maple syrup and dried rose petals, not only to reap the benefits of elderberries' healing powers, but also to indulge in an evening 'brew' that sets you up for a restful, immune-boosting sleep.

# ASHWAGANDHA

In this fantastically soothing moon milk recipe, coconut milk is combined with ashwagandha powder, spices, vanilla and maple syrup for a tonic that tastes like a subtle eggnog.

### Ingredients

500 ml (16 fl oz) unsweetened coconut milk
1 teaspoon ashwagandha powder
¼ teaspoon ground cinnamon
⅛ teaspoon nutmeg
1 tablespoon maple syrup
1 teaspoon pure vanilla extract

The ashwagandha plant is known as an 'adaptogen' and has lots of uses – primarily its ability to help the body cope with daily stress and insomnia. Adaptogens are plants or herbs that are included in your diet to help balance your hormones and adrenal system. They get their name from their ability to adapt to what your body needs, and they regulate your system depending on your need in a given moment. Other research has hinted at ashwagandha being effective in helping to treat anxiety, ADHD, bipolar disorder, diabetes, high cholesterol, male infertility, osteoarthritis, OCD, Parkinson's disease, rheumatoid arthritis and fibromyalgia, among others.

The sweet, nutty aroma and flavour of this milk is both comforting and calming. This recipe will be a winning choice after a long, hard day, to quiet the brain, lower the blood pressure and help your immune system.

# BLUEBERRY LAVENDER

Finish your day with one of the most on-trend flavour combos – blueberry and lavender – swirled with warm, creamy almond milk, maple syrup and raw cacao butter.

**Ingredients**

500 ml (16 fl oz) unsweetened almond milk
½ teaspoon lavender paste (or 1 teaspoon lavender extract)
1 tablespoon blueberry powder
1 tablespoon maple syrup
1 tablespoon raw cacao butter
1 tablespoon dried blueberries, to garnish (optional)

The aroma and flavour of lavender are instantly recognisable, and it is often used in a diffuser or essential oil form for its calming effects. Lavender has become a fundamental part of the medicinal herb garden. It is a mood elevator and antidepressant, and used for the treatment of digestive problems, insomnia and tension headaches.

In this blissful moon milk, the lavender taste is subtle, as lavender can be a little overpowering and may not be the most popular of edible herbs, but when laced with blueberry powder it creates just the right marriage of flavours. Raw cacao butter is also known as a mood improver, creating a feeling of euphoria. Being in a better mood before bed stops your mind from racing, lowers your blood pressure and helps you get a more restful sleep.

# CAKE BATTER

This recipe is a piece of cake. The different ingredients marry together to create the flavour of cake batter in a cup in under five minutes, and boast a wide range of benefits.

### Ingredients

500 ml (16 fl oz) unsweetened hemp milk
1 tablespoon agave syrup
1 teaspoon pure almond extract
1 tablespoon raw cacao butter
½ teaspoon ashwagandha powder
Pinch of salt
1 teaspoon edible glitter stars and pearls (ideally plant-based, gluten-free, vegan), to garnish (optional)
1 tablespoon whipped cacao butter, to garnish (optional)

While moon milk has only recently found a home on our social media accounts, cake batter has been made in our homes for centuries. But satisfying your sweet tooth with a carb- and sugar-heavy dessert before bed has proven to be a factor in poor sleep. So if you find yourself craving a late-night slice of cake, try whipping up this snooze-inducing option instead for sweet dreams.

By adding one of the more popular adaptogens, ashwagandha, you can help to promote calming energy and reduce stress. Hemp milk offers nutritious, healthy fats and a good source of protein, while raw cacao butter is rich in antioxidants and essential fatty acids, which are great for their anti-inflammatory effects and benefits to heart health. Find me a cake that can offer such amazing benefits while at the same time satisfying your sweet tooth. . .

# ACTIVATED CHARCOAL

Activated charcoal is taking the world by storm. This is not the charcoal used to light your barbecue, but a fine black powder made from bone char, coconut shells, olive pits or sawdust, 'activated' by processing it at very high temperatures.

## Ingredients

500 ml (16 fl oz) unsweetened almond milk
1 teaspoon activated charcoal (food-grade)
1 tablespoon maple syrup
1 teaspoon pure vanilla extract
Pinch of salt

## Tip

When purchasing activated charcoal, select powder that is made only from natural sources. It is very important to note that this is not the charcoal used on your barbecue.

Adding this 'superfood' to your diet helps trap toxins, preventing your body from absorbing them. This trendy new powder is used as a teeth whitener in some toothpastes and mouthwashes, and can also be used to treat acne and insect bites, to reduce gas and high cholesterol, and for its anti-ageing properties. It may even prevent a hangover.

In this moon milk recipe, almond milk is heated and whisked with the magical activated charcoal powder, along with maple syrup, vanilla extract and a pinch of salt. Enjoy this warm, super-detoxifying elixir before bed after a long day filled with perhaps one too many boozy beverages, or simply to set your body up for a night of purification while you slumber.

# CRANBERRY ORANGE

Cranberry and orange have been paired together for centuries. The tarty sweetness of the cranberries harmonises with the aromatic citrus for a moon milk made in heaven.

### Ingredients

500 ml (16 fl oz) unsweetened almond milk
¼ cup dried cranberries
1 tablespoon agave syrup
1 teaspoon pure vanilla extract
Juice of ½ orange
1 or 2 orange slices, to garnish (optional)

### Tip

Be sure to check the ingredients on your dried cranberries, as some have a lot of added sugar, which you don't want. Try to choose those that are sweetened with apple juice or are just dried cranberries.

As well as being one of the best flavour combinations of all time, cranberries and orange juice give this moon milk a boost of vitamin C to help fight those nasty colds while you slumber.

You can try this recipe with different varieties of orange, too. There are lots to choose from, all of which offer a fun and unique spin on this moon milk.

# GRAPE

Some believe that the sweetest grapes have a magical ability to turn bad luck into good. Hopefully, this moon milk will bring you the luck of a good night's sleep.

### Ingredients

500 ml (16 fl oz) unsweetened almond milk
1 tablespoon agave syrup
1 teaspoon pure almond extract
4 teaspoons grape powder
1 tablespoon raw cacao butter

### Tip

If you desire a stronger grape flavour, add additional grape powder, one teaspoon at a time.

This magical fruit is loaded with antioxidants and potassium. Potassium helps our muscles to relax as we head in for the night. Grapes are also bursting with nutrients that can help prevent heart disease, constipation and high blood pressure.

Grapes are naturally low in sugar, surprisingly, so indulging in this semi-sweet warm milk before bed can curb that dessert craving without interrupting your sleep. And the phytonutrients found in grapes are now thought to play a role in longevity - perhaps alluding to the magical good luck of those who consume them.

# SPIRULINA

This popular fresh- and salt-water organism is arguably the most nutritious food on the planet. Its deep, rich colour sets it apart from other herbs and supplements, and in this recipe it creates a soothing beverage that is way better than counting sheep.

## Ingredients

500 ml (16 fl oz) unsweetened almond milk
1 tablespoon agave syrup
1 teaspoon spirulina powder
¼ teaspoon ground ginger
¼ teaspoon ground cinnamon

## Tip

I have used aqua spirulina, but you are more than welcome to substitute green spirulina, which will give a similar taste and pack just as much nutritional punch – the only difference will be the colour of your drink.

Did you know that spirulina became even more popular when it was proposed by NASA that it could be grown in space to be consumed by astronauts as a vital source of nutrients? That certainly classifies it as a 'super' superfood.

With the addition of spirulina, this moon milk is particularly high in protein and iron, so if you are on a plant-based diet, this could be a great option if you're looking to top off your protein intake for the day. It's also a good source of potassium and magnesium, both of which aid sleep.

# BANANA CREAM PIE

Just imagine your favourite banana cream pie served warm in a mug in your hands in just five minutes. . .

### Ingredients

500 ml (16 fl oz) unsweetened hazelnut milk
1 teaspoon pure almond extract
1 teaspoon ground cinnamon
1 tablespoon agave syrup
½ banana
Pinch of salt
1 or 2 sugar-free, dried banana slices, to garnish (optional)
1 teaspoon grated almonds, to garnish (optional)

### Tip

Strain the moon milk if desired – if you don't strain, you'll have a thicker milk. It's your preference.

This moon milk has all the taste of a banana cream pie minus all the work, waiting and unwanted calories. Bananas are high in magnesium and potassium, which are known as muscle relaxants to aid deeper sleep. So not only does the moon milk taste like dessert, it's also packed with some pretty great nutrients.

The hazelnut milk gives the recipe a nuttier flavour that pairs well with the bananas, but it's also rich in folic acid, omega-3 fatty acids, B vitamins and vitamin E.

# CHAMOMILE

Instead of your usual chamomile tea, try this deliciously aromatic chamomile moon milk.

### Ingredients

500 ml (16 fl oz) unsweetened almond milk
2 tablespoons date syrup
1 chamomile tea bag
1 tablespoon coconut oil
Pinch of ground ginger
Pinch of ground cardamom
Pinch of ground nutmeg
1 teaspoon dried chamomile buds, to garnish (optional)

Commonly regarded as a mild tranquiliser or sleep inducer, chamomile moon milk is the perfect beverage before bedtime.

In this recipe, I've paired chamomile tea with a pinch of ginger, cardamom and nutmeg. Chamomile tea contains an antioxidant called apigenin, which may help initiate sleep and reduce anxiety. The result is a fragrant, warm beverage to calm your senses and get you some much-needed rest.

# PEPPERMINT

Renowned for its calming properties, try this lightly peppermint-flavoured moon milk.

### Ingredients

500 ml (16 fl oz) unsweetened almond milk
¼ teaspoon pure peppermint extract
1 tablespoon coconut oil
1 tablespoon maple syrup
Mint leaves, to garnish (optional)

Peppermint is a cooling, relaxing herb that helps to calm muscles and cramps, and ease inflamed tissues. This herb has been used for centuries as a remedy and soothing aid for intestinal issues.

Aside from giving an upset stomach some relief, the calming properties of peppermint also help to induce sleep. This moon milk has only a hint of peppermint, so it's not an overwhelming flavour - it's just enough to ignite your senses and help you get some r & r before bedtime.

# BEETROOT ROSE

This moon milk is for beetroot lovers and non-beetroot lovers alike. Add beetroot powder to your milk for a kick of serotonin to help promote sleep.

**Ingredients**

500 ml (16 fl oz) unsweetened coconut milk
½ teaspoon beetroot powder
1 tablespoon maple syrup
1 tablespoon raw cacao butter
1 teaspoon pure vanilla extract
Pinch of salt

Ease into your evening with this rosy brew. Warm coconut milk is blended with beetroot powder and maple syrup for a hint of sweetness. This frothy drink is ideal for including in your bedtime routine.

Serotonin, also known as the 'happy' chemical, plays a role in maintaining mood balance, social behaviour, appetite, digestion, memory, and you guessed it: sleep. Beetroot is also shown to lower blood pressure, helping your body to relax and snooze better.

Beetroot is also high in folate, so be sure to drink up, all you mammas-to-be.

# LAVENDER CHAMOMILE

Both lavender and chamomile are well known for their calming qualities, so this moon milk is sure to have you relaxed before you even close your eyes.

## Ingredients

500 ml (16 fl oz) unsweetened almond milk
1 teaspoon dried lavender buds (food-grade), plus extra to garnish (optional)
2 tablespoons honey (or maple syrup, if vegan)
1 chamomile tea bag
1 teaspoon pure vanilla extract
Pinch of salt

## Tip

Add your chamomile tea bag to the saucepan along with all the other ingredients. You can let your milk simmer a bit longer for a more potent chamomile flavour. You may also wish to strain your milk to remove the lavender buds prior to drinking.

Lavender and chamomile have been used for many years to help people snooze. The sedative effect is caused by apigenin, a nutrient that reduces anxiety, leaving you feeling relaxed and sleepy - perfect for unwinding before bedtime.

The combination of lavender and chamomile also has digestive- and nervous-system-sedating properties, making this lavender chamomile moon milk the perfect after-dinner beverage.

Lavender chamomile tea has been a popular blend over the years, and this milk ups your tea game and ignites your senses using a flavour combination that is very much an old favourite.

# COCONUT CREAM PIE

A thick and creamy, dessert-like moon milk. Ditch the refined sugar and processed ingredients, and opt for a more natural beverage that will fill you up and wind you down.

## Ingredients

500 ml (16 fl oz) unsweetened coconut milk
1 teaspoon pure vanilla extract
1 tablespoon maple syrup
¼ teaspoon ground cinnamon
Pinch of salt
Toasted shredded coconut, to garnish (optional; see below)

## Tip

To toast your shredded coconut, place about 4 tablespoons in a small frying pan over a medium-high heat until it starts to turn golden brown, tossing frequently to avoid burning.

Coconut milk has become popular for its creamy texture when added to soups, curries, oatmeal, lattes and many other dishes, but did you know that it has a pretty great reputation for building up the body's immune defences to help prevent sickness and disease? It's also a good source of magnesium, helping to combat anxiety, stress and muscle tension, and aiding relaxation so you can get to sleep faster.

In this warm beverage, coconut milk is blended with one of the most popular additions to moon milk: cinnamon. It has a strong yet sweet flavour and gives an inviting aroma to literally everything you add it to.

# BLACK SEED

Black seed is a plant that has been used in medicine for over 2,000 years – it was even discovered in the tomb of King Tut. In this moon milk recipe, the ancient seed gets a modern-day upgrade to help you catch up on your beauty sleep.

**Ingredients**

- 500 ml (16 fl oz) unsweetened almond milk
- 1 teaspoon black seed powder
- ¼ teaspoon ashwagandha powder
- ¼ teaspoon ground ginger
- 1 tablespoon maple syrup

This recipe combines almond milk, black seed powder, ashwagandha powder, ginger and maple syrup to help boost your mood and memory, all the while offering stress relief.

Most people haven't heard of black seeds, but they are quickly gaining in popularity for all the benefits they provide. You may have heard of or used the culinary spice cumin; black seed is actually referred to as black cumin, and is a promising component effective in fighting 'superbugs' - strains of bacteria and viruses that are antimicrobial-resistant and difficult to treat.

With more research finding many other uses for black seed, why not whip up a cup of this moon milk next time influenza strikes to help kick the virus faster. And with the inclusion of adaptogenic ashwagandha powder, it is certainly the one to beat for an amazing night's sleep.

# RED VELVET

Traditionally, red velvet cake had a buttermilk or vinegar component that made it super velvety, while the colour came from cocoa powder or food colouring. The red colour in this milk comes from the more natural source of beetroot powder.

**Ingredients**

500 ml (16 fl oz) unsweetened oat milk
1 tablespoon beetroot powder
1 teaspoon apple cider vinegar
1 teaspoon pure vanilla extract
1 tablespoon raw cacao butter
1 tablespoon agave syrup

This moon milk recipe tastes just like the popular dessert. In vegan baking, a 'buttermilk'-like mixture is achieved by combining plant milk and apple cider vinegar, and letting it curdle. In this recipe, for time's sake, we don't let it curdle, but we do use oat milk combined with apple cider vinegar and raw cacao butter to give the velvety rich flavour - and add some nutritional benefit too.

Apple cider vinegar has been gaining attention for its amazing health benefits. It's known to help in lowering blood sugar levels and cholesterol, and relieving acid reflux. It also contains antioxidants and is believed to prevent premature ageing. The beetroot powder lowers blood pressure, to help you get that extra deep sleep you're longing for.

# BLUE BUTTERFLY PEA FLOWER

The qualities of depth, trust, loyalty, sincerity, wisdom, confidence, stability, faith, heaven and intelligence are all represented by the colour blue.

**Ingredients**

500 ml (16 fl oz) unsweetened almond milk
1 teaspoon blue butterfly pea flower powder
1 teaspoon pure almond extract
Pinch of salt

Sadly, butterfly pea flowers have a fleetingly short life span of only twenty-four hours before they wither. Fortunately, dried butterfly pea flowers last a little longer and can be used as a natural dye while offering eyesight improvement, hair growth and skin health.

Indulge in this beautiful blue moon milk for a positive end to your day, both mentally and physically.

And if you'd like to experiment a little, the milk will change colour when the pH balance changes - it turns purple with the addition of lemon juice.

# TURMERIC

You may have heard that turmeric is packed with vitamins and minerals and possesses anti-inflammatory properties, but it is also high in protein and iron, and has been shown to be effective in treating depression. That's a lot of amazing advantages to adding this magical golden spice to your diet.

**Ingredients**

500 ml (16 fl oz) unsweetened almond milk
1 tablespoon raw cacao butter
½ teaspoon ground turmeric
¼ teaspoon ground cinnamon
1 tablespoon maple syrup
1 teaspoon pure vanilla extract
Pinch of pepper and salt

In this more savoury moon milk, with turmeric at its base, warm almond milk is combined with raw cacao butter, cinnamon, maple syrup and pure vanilla extract for a liquid- gold potion that lowers your blood sugar levels, reduces inflammation and eases your digestive system, helping you relax before bedtime. It also works its magic in helping your liver to detoxify, boosting your immune system while you sleep.

# WHITE CHOCOLATE

Perfect for lovers of white chocolate. This is one creamy and dreamy moon milk to help get you to sleep.

### Ingredients

- 500 ml (16 fl oz) unsweetened almond milk
- 2 tablespoons raw cacao butter
- 1 tablespoon maple syrup
- ¼ teaspoon pure almond extract
- 1 teaspoon dried rose petals (food-grade), to garnish (optional)

While there is no actual white chocolate in this recipe, raw cacao butter replaces the sugary alternative. And paired with almond extract, it tastes similar to the white chocolate you know and love.

We know cacao butter as chocolate, but when it is in its less processed form, it's full of healthy fats, nutrients and antioxidants, and helps to balance your mood. It's the ideal moon milk to feed your body before it rejuvenates overnight.

# ROASTED DANDELION

Think twice about getting rid of those pesky yellow 'weeds' that take over your garden in the warmer months, because they actually have some pretty amazing health benefits.

### Ingredients

500 ml (16 fl oz) unsweetened almond milk
1 roasted dandelion tea bag
1 tablespoon raw cacao butter
1 tablespoon agave syrup
Pinch of salt
1 tablespoon roasted dandelion (food-grade), to garnish (optional)

### Tip

The longer you let the tea bag simmer in the milk, the stronger the dandelion taste will be.

Dandelions have been used for centuries for their medicinal properties. Whether you add fresh dandelion leaves to your salad, cook them like spinach, or use the whole plant dried, roasted and ground up in your tea (or moon milk), this mildly bitter ingredient helps your body's natural detoxification process. Dandelion also lowers your blood pressure and is a good source of potassium, helping you to relax.

Why not curl up with a cup of this roasted dandelion moon milk and let it get to work while you rest?

# PECAN PIE

This moon milk basically gives you a relaxing massage while satisfying that dessert craving.

### Ingredients

500 ml (16 fl oz) pecan milk (see below)
½ teaspoon pure vanilla extract
1 tablespoon maple syrup
Pinch of ground cinnamon, to garnish
1 or 2 raw pecans, to garnish (optional)

**For Homemade Pecan Milk**

200 g (7 oz) raw pecans
3.3 L (7 pints) cups purified/ filtered water
6 dates
1 teaspoon pure vanilla extract
½ teaspoon salt

Pecans help to regulate your heart rhythm and promote relaxation, kind of like an internal massage which helps you to fall asleep faster. They are also a great source of tryptophan, which is used to make the hormone melatonin, a known contributor to a good night's sleep.

### HOMEMADE PECAN MILK

Soak the pecans in 1.4 L (3½ pints) of purified/filtered water overnight (I keep mine in the fridge). The next day, drain and rinse your pecans, then place them in a blender with the dates, vanilla, salt and the remaining water. Blend on high until the mixture is finely blended (about a minute). Position your nut milk bag (or muslin) over a large bowl or jug and pour the nut milk purée into the bag. Twist and knead the bag gently, allowing the filtered milk to collect until the pulp is dry. Refrigerate and serve chilled, or use in this moon milk recipe. You can also repurpose the pulp and use it to make pastry or muesli.

# UNICORN

The health benefits of unicorn moon milk are just as magical as the mysterious creature after which it is named.

## Ingredients

500 ml (16 fl oz) unsweetened almond milk
1 teaspoon coconut oil
2 teaspoons maple syrup
1 tablespoon pink pitaya pea flower powder (for the pink moon milk)
1 tablespoon blue butterfly pea flower powder (for the blue moon milk)
1 teaspoon emerald pandan leaf powder (for the green moon milk)

Combine these three moon milk 'shots' to harvest the benefits of three magical plant-based powders.

The pink moon milk is coloured with pitaya (dragonfruit) powder. This gorgeous pink powder helps boost the immune system with its many antioxidants and vitamin C, while also boosting the metabolism and soothing the digestive system. This superfruit also contains several types of vitamin B, protein, fibre and essential fatty acids.

Butterfly pea flowers are found across Thailand and Myanmar, and have a life span of just twenty-four hours. The dried butterfly pea flower, however, can be used as a natural dye and also helps to maintain healthy skin and hair, and improve eyesight.

Pandan leaf powder is often used in Asian cuisine to bring a vanilla-like aroma to dishes. It is a beautiful, bold emerald colour and can provide relief from arthritis, headaches, fever and skin problems.

# MAPLE ALMOND

End the day with some deep breathing and the soothing aroma of maple and almond that drifts to your nose with each sip.

**Ingredients**

500 ml (16 fl oz) unsweetened almond milk
2 tablespoons maple syrup
½ teaspoon ashwagandha powder
¼ teaspoon pure almond extract

While not as colourful as some of the other creative concoctions in this book, don't write off this moon milk just yet. Bringing it back to basics never tasted so sweet - naturally.

This recipe is the fusion of warm almond milk with maple syrup, ashwagandha powder and almond extract. Maple syrup is one of the most popular natural sweeteners worldwide as an alternative to cane sugar, and when consumed in moderation it actually has some surprising health benefits. It contains numerous antioxidants - up to twenty-four, actually - helping to protect your skin, and also fighting inflammation, cancer and neurodegenerative diseases.

When paired with ashwagandha powder, a substance that enhances the body's adaptive response to stress and balances normal body functions, this moon milk moves from underdog to an easy front-runner.

# CHERRY PIE

'An apple a day keeps the doctor away', but how about using tart cherry juice to keep insomnia at bay.

### Ingredients

500 ml (16 fl oz) unsweetened macadamia milk
¼ cup tart cherry juice
1 tablespoon coconut oil
1 tablespoon agave syrup
1 teaspoon pure almond extract
Pinch of ground nutmeg
1 or 2 fresh tart cherries, halved, to garnish (optional)

### Tip

Make sure to use tart cherry juice rather than regular cherry juice, as only tart cherries contain melatonin, used for sleep regulation. Also look for tart cherry juices that are not from concentrate and with no added sugars.

Tart cherries contain large amounts of melatonin, the sleep hormone. Drinking it before bed can not only get you to sleep faster, but studies also show you will sleep for longer.

Tart cherry juice also offers other benefits besides quality sleep, including better muscle recovery and reduced joint pain and inflammation. Macadamia milk is high in magnesium and potassium, helping you relax and fall asleep quicker, so basically this is one heck of a sleep tonic.

# CARAMEL

Caramel has been used for centuries in the finest bakeries. Add caramel-flavoured moon milk to your list of delicate 'desserts'.

**Ingredients**

500 ml (16 fl oz) unsweetened almond milk
1 tablespoon maple syrup
2 tablespoons almond butter
1 tablespoon coconut oil
Pinch of salt

This moon milk has a subtle caramel taste without using any 'real' caramel. The almond milk is combined with maple syrup, almond butter, coconut oil and a pinch of salt to remind you of your favourite sweet condiment without being loaded with sugar.

The almond butter used in this recipe provides some healthy fats, vitamin E, and protein that offer many health benefits; it is also high in magnesium and potassium, helping to calm your muscles and leading to a more restful sleep.

# PITAYA ALOE

Impress your friends on social media with this pretty pink drink. Combine warm almond milk with pitaya powder, aloe powder, vanilla and cinnamon for a swoon-worthy sleep tonic.

### Ingredients

500 ml (16 fl oz) unsweetened almond milk
1 teaspoon aloe vera powder
¾ teaspoon pink pitaya pea flower powder
⅛ teaspoon ground cinnamon
1 tablespoon maple syrup
1 teaspoon pure vanilla extract

Pitaya - also known as dragonfruit - is used around the world in many different dishes, not only for its beauty, but for its health benefits, too. It's comprised of antioxidants and B vitamins, and has antibacterial properties, resulting in a boosted immune system and calmer digestion. When paired with aloe in this recipe, it's a match made in heaven.

Aloe vera makes its biggest appearance during the summer as a remedy for sunburn, but in edible form it's consumed for digestive disorders and inflammation of the stomach. A calmer tummy equals better sleep, so you wake up feeling refreshed and ready to tackle the day.

# CARDAMOM

Big things come in tiny packages, and that is undoubtedly true of cardamom seeds. Cardamom is commonly used in Ayurvedic practices, making this moon milk pretty much the definition of a sleepy tonic. And it possesses aphrodisiac properties too. . .

**Ingredients**

500 ml (16 fl oz) unsweetened hemp milk
1 tablespoon agave syrup
½ teaspoon ground cardamom
½ teaspoon ashwagandha powder

Cardamom has been used to treat various gastrointestinal issues, to control cholesterol and relieve cardiovascular issues, to improve blood circulation, and cure dental diseases and urinary tract infections, to name just a few. When paired with ashwagandha powder, a de-stressing superherb, you get the perfect pairing: snug in a mug.

Cardamom can be used in a lot of recipes in seed form and is harvested as such, but this recipe calls for ground cardamom. Although cardamom can be expensive, you'll find a little goes a long way, so a single jar will allow you to make this recipe again and again.

# STRAWBERRY SCONE

Swap out sugary, processed desserts with this naturally sweetened moon milk. You'll find this recipe is packed with good fats and nutrients, a perfect pre-bedtime treat.

## Ingredients

500 ml (16 fl oz) unsweetened coconut milk
1 teaspoon pure vanilla extract
⅛ teaspoon ground cardamom
1 tablespoon agave syrup
½ cup freeze-dried strawberries, plus extra to garnish (optional)

## Tip

For a lower fat option, substitute full-fat coconut milk with light coconut milk. You can also use maple syrup in place of agave syrup, if desired. Blend your milk after heating to grind up those freeze-dried strawberries.

If you're the type who loves to eat something sweet as you're plopped on the couch in the evening, this dessert-like moon milk is for you. It tastes like a strawberry scone and will leave you feeling relaxed and satisfied as you head to bed for the night.

Fresh strawberries are a seasonal fruit, so freeze-dried strawberries offer the same great taste while preserving nutritional value. Strawberries contain vitamin C and antioxidants, have been shown to reduce inflammation and they are even thought to be an aphrodisiac.

# GINGER TURMERIC

Unlike many of the other moon milks in this book, this recipe steers away from a sweet, dessert-like flavour, providing a stimulating kick from the turmeric, ginger and pepper.

## Ingredients

500 ml (16 fl oz) unsweetened cashew milk
½ teaspoon ground ginger
½ teaspoon ground turmeric
¼ teaspoon ground cardamom
½ teaspoon ground cinnamon
1 tablespoon maple syrup
1 teaspoon pure vanilla extract
Pinch of pepper and salt

Sip on this vibrant moon milk after a long day of indulging in too many 'cheat' foods to give your body some valuable spices, helping you settle in for the night.

Both ginger and turmeric have been used for centuries for their known positive effects on the digestive system, and the combination of them in this moon milk is one unbeatable marriage that helps you battle nausea and indigestion. Cardamom also boosts digestive health and helps to lower blood pressure, allowing your body to relax before you hit the hay.

# WHITE CHOCOLATE RASPBERRY

Warm coconut milk is blended with raw cacao butter and raspberries for a frothy, relaxing, scrumptious bedtime delight.

### Ingredients

500 ml (16 fl oz) unsweetened coconut milk
¼ cup freeze-dried raspberries, plus extra to garnish (optional)
1 tablespoon maple syrup
1 teaspoon pure vanilla extract
1 tablespoon raw cacao butter

### Tip

Strain the warm milk if you prefer a smoother, less seedy texture.

Other than being nature's candy, these rich-coloured berries contain a variety of vitamins, minerals and antioxidants. Foods high in vitamin C, such as red raspberries, may even protect the eyes from sun damage, filtering out harmful blue light rays.

Revitalise your eyes in more ways than one when you indulge in this fruity, creamy beverage before getting some shut-eye. The combination of the raspberries with the raw cacao butter means you're bound to wake up feeling well rested, rejuvenated and - with the anti-inflammatory and anti-ageing effects of this moon milk - maybe even a tad younger.

# CHERRY THYME

Thyme is so much more than just a pretty garnish. The distinctive taste of this primeval herb – a kin to its popular cousin, mint – makes it a culinary staple in our kitchens, and it is also becoming increasingly known for its medicinal qualities.

### Ingredients

500 ml (16 fl oz) unsweetened oat milk
¼ cup tart cherry juice
5 large sprigs of fresh thyme
1 teaspoon pure almond extract
1 tablespoon agave syrup
Pinch of cinnamon
Pinch of pepper

### Tip

If you prefer a stronger thyme flavour, simply let the sprigs steep in your milk mixture before removing. And make sure to use tart cherry juice rather than regular cherry juice, as only tart cherries contain melatonin, used for sleep regulation.

This cherry thyme moon milk feels like a mischievous indulgence, but unlike some cherry-flavoured beverages which are high in sugar, you can sip your way to better immunity, a better mood and better blood pressure with this low-sugar alternative.

Thyme has the ability to help treat acne with its antibacterial properties, and is also effective in lowering blood pressure, which can ultimately help your body relax and sleep better. And tart cherry juice is nature's gift to us as one of the highest natural sources of melatonin, a known component in achieving quality sleep.

# GOLDEN MILK

This bright yellow moon milk has been gaining in popularity, and for good reason – it is hailed for its immune-boosting properties, as well as its ability to reduce inflammation and improve mood and digestion.

**Ingredients**

500 ml (16 fl oz) unsweetened almond milk
¼ teaspoon ground cardamom
½ teaspoon ground turmeric
¼ teaspoon ground cinnamon
⅛ teaspoon ground ginger
1 tablespoon maple syrup
1 tablespoon raw cacao butter
Pinch of salt

With the combination of cardamom, turmeric, cinnamon and ginger, this boldly spiced recipe not only helps you sleep like a baby, but also combats any nasty bugs as you do so, making this a win-win night-time treat.

Heat a pot of this on the stove, slip into your comfiest PJs, put on some relaxing music and let this golden milk work its magic while you slumber.

# LAVENDER

Perhaps the ultimate sleep tonic, this moon milk offers more than just its gorgeousness to your bedtime routine.

## Ingredients

500 ml (16 fl oz) unsweetened almond milk
1 teaspoon pure vanilla extract
½ tablespoon dried lavender buds (food-grade), plus extra to garnish (optional)

## Tip

If you prefer a smoother texture, you can use a coffee grinder to grind your lavender buds very finely, or alternatively you can strain your moon milk before consuming.

Lavender has been used for many years for its medicinal benefits in treating anxiety, restlessness and insomnia.

You may think that lavender is used solely as a scent in candles, perfumes or essential oils, but did you know you can ingest it as well? When consumed, it has been shown to help with digestive issues.

Combining lavender with almond milk in this recipe will ensure you turn into bed early and keep those zzz's going all night long.

C

# CINNAMOON MUSCLE MILK

Enhance your muscular recovery and development while you sleep after drinking this fragrant, protein-packed recipe.

## Ingredients

500 ml (16 fl oz) unsweetened hemp milk
1 tablespoon maple syrup
½ tablespoon vanilla protein powder
1 teaspoon ground cinnamon
Pinch of salt

## Tip

Too much sugar before bed can have an adverse effect on sleep. Be sure to check that your protein powder is low in sugar.

Sleep is an essential component of recovery, not only for our brains, but for all parts of the body, including our muscles. The plant-based protein powder and hemp milk in this moon milk recipe give your body the boost it needs to help with a faster recovery (such as after an intense workout), while calming the mind for a more restful sleep.

Cinnamon's sweet, warming taste has made it one of the world's most loved spices. As little as half a teaspoon a day can have a positive effect on blood sugar, digestion, immunity and heart disease. It can be added to so many different recipes that it's no wonder it's a favourite, and now we have yet another reason to love it.

# PANDAN LEAF

A warm, nutty, vanilla aroma helps settle your body and mind as you slip into your PJs and sip on this vibrant moon milk.

## Ingredients

500 ml (16 fl oz) unsweetened almond milk
1 teaspoon pandan leaf powder
1 tablespoon maple syrup
1 teaspoon pure vanilla extract

Aside from their gorgeous emerald hue, pandan leaves offer many health benefits, including relief of headaches, fevers, skin problems, oral discomfort and arthritis pain.

Incorporating this warm pandan leaf tonic into your bedtime routine is sure to help you fall asleep faster, and with all those healing properties your quality of sleep will be much improved.

# PINK PITAYA

Besides looking absolutely gorgeous, the pink pitaya is an ancient Ayurvedic drink used to help with sleeplessness.

### Ingredients

500 ml (16 fl oz) unsweetened almond milk
1 teaspoon coconut oil
2 teaspoons maple syrup
1/2 teaspoon ground cinnamon
1 teaspoon pink pitaya pea flower powder
Pinch of ground nutmeg
Pinch of ground cardamom
Dried rose petals (food-grade), to garnish (optional)

### Tip

If you are using rose petals, which I highly recommend, but would prefer a smoother texture, try grinding the petals into finer pieces before garnishing your milk.

The idea of drinking warm milk before bed to help you sleep stems from this Ayurvedic remedy. Here, I am giving this ancient moon milk a hip new twist by adding colourful superfoods and fragrant spices. So not only will it help with the occasional bout of sleeplessness, but by amping up the natural spices it will also boost your immune system. In other words, it rocks.

And as I always say, everything is better pink, so I decided to add pitaya (dragonfruit) powder and dried rose petals to mine, along with cinnamon, nutmeg and cardamom.

# CHAI

Known as a way of life in India, chai is everywhere you look. It's sweet and spicy with cinnamon or cardamom notes and milky undertones. Next time you have a hankering for a soothing chai latte, consider making your own moon milk version at home.

**Ingredients**

500 ml (16 fl oz) unsweetened hazelnut milk
1 chai tea bag
1 teaspoon coconut oil
½ teaspoon ashwagandha powder
1 tablespoon maple syrup
1 teaspoon pure vanilla extract
1 teaspoon chai spices, such as cloves and ginger, to garnish (optional)

This chai moon milk encompasses the basic components of chai tea: tea, milk, spices and sweetener. To keep this recipe quick and simple, use your favourite chai tea bag; the base is usually a black tea, with a number of different spices, such as cardamom, cinnamon, ginger, star anise and cloves. Chai almost always includes milk - in this case, hazelnut - to give it a robust, nutty flavour. And to really bring out the flavour of the spices, chai typically involves a teaspoon of sugar, but in this recipe maple syrup is used instead to keep the sugar content low.

This recipe also calls for half a teaspoon of ashwagandha powder, known as an adaptogen, which helps to reduce stress and aid sleep.

# BLUE MOON MILK

Your new favourite way to enjoy a Blue Moon. This moon milk combines the superfood blue spirulina with a splash of orange juice in warm, creamy coconut milk to help you settle in as the moon rises.

**Ingredients**

500 ml (16 fl oz) unsweetened coconut milk
1 tablespoon maple syrup
1 teaspoon pure vanilla extract
½ teaspoon blue spirulina powder
1 tablespoon orange juice

You may have heard the saying 'once in a blue moon', signifying the rarity of the 'extra' full moon seen every few years. Like the astronomical anomaly, this recipe is exceptional in both its flavour combination and appearance. Although the term 'blue moon' isn't a literal interpretation of the colour of the moon, the gorgeous blue hue of this dreamy drink is something to savour.

Blue spirulina is harvested from blue-green algae. The vibrant blue powder is gaining popularity for its richness in antioxidants and health benefits, helping to lower blood pressure and blood sugar levels, aiding a more restful sleep.

# HAZELNUT

Keep it simple with this warm and comforting hazelnut moon milk.

## Ingredients

- 500 ml (16 fl oz) unsweetened hazelnut milk
- 1 tablespoon maple syrup
- ½ teaspoon ground cinnamon
- ¼ teaspoon pure vanilla extract
- 1 teaspoon grated hazelnut, to garnish (optional)

Hazelnuts keep your blood pressure in check and boost your heart health. They're also good for your skin, as they're a great source of vitamin E.

With this moon milk containing only four ingredients, you can heat it up in no time. Hazelnut milk is extra creamy, making it a truly indulgent moon milk.

# CHERRY COCOA

Think hot cocoa with a hint of cherry.

### Ingredients

500 ml (16 fl oz) unsweetened coconut milk
1 tablespoon date syrup (or maple syrup)
¼ cup tart cherry juice
1 teaspoon unsweetened cocoa powder
Pinch of salt

Studies have found that incorporating tart cherries into your daily diet can help regulate your sleep patterns. So when you're sipping on this cup of healthy hot cocoa, you're actually sipping on a cup of sleepy serum.

Cocoa powder has a bad reputation for its high caffeine content, but in small amounts, such as in this recipe, it has some really great health benefits, including reducing inflammation, improving heart and brain health, and blood sugar control, without the side effects of too much caffeine.

# BLUEBERRY PIE

This moon milk recipe is as easy as pie. It's a known fact that blueberries are armed with antioxidants, vitamins and minerals – so why not add this powerhouse to your nightly ritual?

## Ingredients

500 ml (16 fl oz) unsweetened hazelnut milk
4 teaspoons blueberry powder
½ teaspoon ground cardamom
¼ teaspoon ground cinnamon
1 tablespoon maple syrup
1 teaspoon pure vanilla extract

Everyone likes pie, don't they? So imagine the taste of your favourite warm blueberry pie as a drink, made in just five minutes, with only six ingredients, and a whole host of beneficial effects - and all minus the guilt of indulging in too much sugar and carbs. This moon milk is subtly sweet and can really give your body a healthy boost, while keeping the kitchen clean-up to a minimum.

This moon milk recipe uses hazelnut milk, which, as well as giving your 'pie' a stronger, nuttier flavour, is high in B vitamins that are essential for mental health, helping your brain relax and setting you up for a more peaceful sleep.

# PISTACHIO LEMON

Pistachios contain the antioxidant resveratrol, also found in red wine – but unlike its alcoholic counterpart, this zingy moon milk should keep you hydrated and send you to sleep with no regrets.

### Ingredients

500 ml (16 fl oz) unsweetened pistachio milk
½ tablespoon dried lemon peel
½ teaspoon pure vanilla extract
2 tablespoons agave syrup
¼ teaspoon ground ginger
¼ teaspoon ashwagandha powder
¼ teaspoon ground cinnamon
1 or 2 unwaxed lemon slices, to garnish (optional)

***For Homemade Pistachio Milk***
130 g (4¾ oz) shelled pistachio nuts
1.7 L (3½ pints) purified/filtered water

### Tip

You can use the pulp left over from making the pistachio milk to make pastry or muesli.

Pistachios are notably high in B6 vitamins, which are important for blood sugar regulation and the formation of haemoglobin, the molecule that carries oxygen to your red blood cells. Pistachios are also rich in potassium, which can help you to achieve a deeper sleep.

The lemon peel used in this recipe is high in calcium and potassium, which also contributes to a better quality of sleep, while giving you a boost of vitamin C to help fight any oncoming nasties.

## HOMEMADE PISTACHIO MILK

Soak the pistachios in 700 ml (1½ pints) of purified/filtered water overnight (I keep mine in the fridge). The next day, drain and rinse your pistachios, then place them in a blender with the remaining water. Blend on high until the mixture is finely blended (about a minute). Position your nut milk bag (or muslin) over a large bowl or jug and pour the nut milk purée into the bag. Twist and knead the bag gently, allowing the filtered milk to collect until the pulp is dry. Refrigerate and serve chilled, or use in this moon milk recipe.

# STAR ANISE

A flavourful moon milk that will alleviate indigestion and cold symptoms, helping you to get a more restful sleep.

### Ingredients

500 ml (16 fl oz) unsweetened macadamia milk
½ teaspoon ground cinnamon
¼ teaspoon ground ginger
1 tablespoon maple syrup
4 large star anise

### Tip

Star anise has a pretty strong flavour. Remove the pods before consuming if you would prefer a milder milk.

The age-old spice star anise has been used in Asian cuisine for many years, not only for its licorice-like flavouring in both sweet and savoury dishes, but also for its medicinal properties in treating coughs, flu and an upset stomach.

The ginger used in this recipe will also aid in alleviating indigestion. Macadamia milk is high in magnesium and potassium, helping you relax and fall asleep quicker.

# COTTON CANDY

This moon milk looks like candy floss, and tastes like candy floss – but don't be fooled, it's actually good for you.

### Ingredients

500 ml (16 fl oz) unsweetened almond milk
1 teaspoon pure almond extract
1 tablespoon maple syrup
1 tablespoon raw cacao butter
½ teaspoon pink pitaya pea flower powder

### Tip

Be sure to get raw cacao butter, as it hasn't been heated like cacao butter, which destroys any nutrients or vitamins.

Candy floss is straight-up sugar, the opposite of what would help us achieve a great night's sleep - but it's so much fun. This recipe ditches the sugar without losing any of the joy, and proves to be a great indulgence before bedtime.

By using almond milk and almond extract in this recipe, you're getting a good dose of magnesium, which is known to aid sleep. The raw cacao butter supplies your body with essential fatty acids and antioxidants, helping to boost immunity and reduce inflammation.

# PEACH COBBLER

A naturally sweetened beverage that tastes like the real deal and is loaded with nutritional benefits. Satisfy your late-night sweet tooth and head to bed with peachy dreams.

## Ingredients

500 ml (16 fl oz) unsweetened coconut milk
½ teaspoon ground cinnamon
¼ teaspoon ground nutmeg
2 tablespoons agave syrup
1 teaspoon pure vanilla extract
1 ripe peach, peeled and diced

## Tip

When you blend your fresh peach with the milk mixture, it may be rather thick. If you don't mind that, sip away. Alternatively, you can choose to strain your milk mixture before drinking for a smoother texture.

Sip on this vegan moon milk that blends cinnamon and nutmeg with coconut milk and a fresh peach. It is loaded with magnesium and antioxidants – essential elements used by nearly every component of the body.

# BEETROOT

Nothing 'beets' a warming moon milk loaded with nutritional value that can also help you get to destination slumberland.

**Ingredients**

500 ml (16 fl oz) unsweetened almond milk
2 teaspoons beetroot powder
1 tablespoon agave syrup
1 teaspoon pure almond extract
½ teaspoon ground cinnamon
Dried rose petals (food-grade), to garnish (optional)

Beetroot never quite gets the recognition it deserves. As one of nature's most vibrant veggies, not only does the deep red-purple hue of this rooted plant make it a must-have in your diet, but the health benefits are also noteworthy. Beetroot helps to lower blood pressure and increase blood flow, allowing your body to relax, detox and restore.

# MAPLE TAHINI

Maple syrup and tahini make a great flavour combination, so it's no surprise that this moon milk is scrumptious, setting you up for a restful sleep while igniting your senses.

## Ingredients

500 ml (16 fl oz) unsweetened almond milk
2 tablespoons tahini (see below)
1–2 tablespoons maple syrup (see below)
1 teaspoon pure vanilla extract

## Tip

This moon milk is quite thick, so if you desire a thinner milk, try straining the mixture before enjoying, or use only one tablespoon of tahini instead of two. You could also use just one tablespoon of maple syrup if you want to cut down on the sweetness.

Tahini is made from ground sesame seeds, and many people know it only as a vital ingredient in houmous. Tahini is a huge staple in the Middle East, not just because it's extremely versatile and delicious. This butter-like paste is high in protein and healthy fats, and has been shown to lower cholesterol, balance hormones, lower blood pressure and prevent inflammation, among other things, all of which can set your body up for a more restful and restorative sleep.

The use of maple syrup as a sweetener is a low-calorie, natural option, and it also has antioxidant properties that protect the body from free radicals.

# CARROT CAKE

All your favourite carrot-cake flavours in one convenient, cozy cup – now you can have your cake and drink it, too.

## Ingredients

500 ml (16 fl oz) unsweetened oat milk
1 tablespoon carrot powder
¼ teaspoon ground cinnamon
¼ teaspoon pure vanilla extract
1 tablespoon agave syrup
Crushed walnuts, to garnish (optional)

## Tip

If you are having trouble finding oat milk, you can always substitute with almond milk.

Boosting your consumption of the nutrient alpha-carotene, found in carrots, has been closely linked with better sleep. Low intake of this nutrient, on the other hand, has been associated with trouble falling asleep. Carrots are the most potent source of alpha-carotene, behind canned pumpkin, but if you're not a huge fan, disguising them in a warm, delicious beverage may be the best way to add them to your diet.

Growing up, your parents might have told you to eat carrots because they are good for your eyes, and in fact they were right. Vitamin A and the prominent antioxidant lutein, also found in carrots, is especially beneficial for eye health. Other benefits of carrots include reduced risk of cancer and cardiovascular disease, and lower blood cholesterol.

# HIBISCUS FUNFETTI

This cheerful moon milk sings with colour. The tart, berry-like flavour of hibiscus flowers is a firm favourite if you're looking for a naturally sweet option as you settle in for the night.

**Ingredients**

500 ml (16 fl oz) unsweetened cashew milk
1 teaspoon hibiscus flower powder
1 tablespoon raw cacao butter
1 tablespoon maple syrup
1 teaspoon pure almond extract
Sprinkles, to garnish (ideally plant-based, gluten-free, vegan)

Aside from their stunning fuchsia hue, the many health benefits of hibiscus flowers include lowering bad cholesterol and blood pressure, increasing good cholesterol, managing inflammation and improving digestion – all of which help your body relax and fall into a more restful sleep.

After blending the ingredients, jazz up the moon milk with some colourful sprinkles of your choice.

# VANILLA

Aside from having a dreamy scent, did you know that vanilla in its pure form is linked to the hormones that help you drift off?

## Ingredients

500 ml (16 fl oz) unsweetened oat milk
½ teaspoon pure vanilla extract
1 tablespoon agave syrup

Historically, vanilla was used to make a drink fit only for kings to consume. Still a rather royal indulgence in its purest form, it is imperative that you use only pure vanilla extract as it offers the most benefits. Over time, vanilla has become a staple in baked goods for its sweet, notable aroma, and is often also used in lotions, sprays, candles and essential oils.

The aroma that floats up your nose as you take each sip helps to alleviate anxiety and stimulate an overall peaceful and pleasurable feeling, and the indulgent taste supports better sleeping habits. Pure, natural vanilla is also known to be a powerful antioxidant, antibacterial and anti-inflammatory.

So slip into those PJs, light your favourite vanilla-scented candle and sip on your new favourite vanilla 'latte' to transport you quickly to dreamland.

# POMEGRANATE

Known for their booming antioxidants, we should make a real effort to include pomegranates in our diet. This moon milk recipe keeps things simple and uses the juice of this delectable fruit.

### Ingredients

500 ml (16 fl oz) unsweetened oat milk
¼ cup pure pomegranate juice
¼ teaspoon ground ginger
1 tablespoon agave syrup

The name pomegranate derives from the Latin words *pōmum* and *grānātum*, meaning 'seeded apple'. This seedy fruit should earn a spot right up there with kale as one of the healthiest foods. Pomegranates are useful for blood pressure maintenance, improving joint pain and enhancing digestion, and are especially useful in helping you sleep at night.

With a high magnesium content, you may find that pomegrantes are a real superfood for sleep. Diets rich in magnesium have been linked with improved quality, duration and tranquility of sleep. Pomegranates have also been shown to help regulate the metabolism, which in turn reduces sleep disorders and the occurrence of insomnia.

# PUMPKIN SPICE

There's something magical about the pumpkin pie flavour, so if you're counting down the days until pumpkin spice is back on the menu at your favourite coffee shop, this moon milk is for you.

### Ingredients

500 ml (16 fl oz) unsweetened oat milk
1 teaspoon pumpkin spice
1 teaspoon pure vanilla extract
1 tablespoon maple syrup
½ teaspoon ashwagandha powder

### Tip

There may be variations in store-bought pumpkin spice (also known as pumpkin pie spice) and the taste may vary slightly, but all will be delicious. If you cannot find oat milk, you may substitute your favourite plant-based milk with similar results.

Buried beneath pumpkin spice's insane popularity, the blend does offer some great benefits too. Cinnamon, the star ingredient, is rich in antioxidants and aids blood sugar control. Nutmeg contains numerous B vitamins and small amounts of fibre. Ginger is high in iron, potassium and zinc. And cloves also help to stabilise blood sugar levels and additionally maintain brain function with their high composition of manganese. Furthermore, they are known to block the growth of bacteria.

With the addition of ashwagandha powder, mainly known for its ability to help the body cope with daily stress and insomnia, you'll be ditching the caffeinated version and subbing this warm, sleep-inducing beverage in its place.

# CHERRY OATMEAL

You may have had breakfast for dinner, but what about breakfast for dessert? This cherry oatmeal moon milk is packed with sleep-inducing, health-aiding benefits.

**Ingredients**

500 ml (16 fl oz) unsweetened oat milk
¼ cup tart cherry juice
1 tablespoon raw cacao butter
1 teaspoon pure vanilla extract
1 tablespoon maple syrup
Pinch of salt
1 teaspoon oats, to garnish (optional)

While there are no whole oats in the main recipe, this moon milk is made with oat milk, which contains all the benefits of the fibrous grain but in liquid form. Oats are a valuable source of plant-based protein, vitamins, and minerals, and are rich in antioxidants, helping to lower blood sugar and blood pressure, making your body relax and sleep better. Oats can also help you feel full for longer, prolonging your night hibernation.

Researchers have found that tart cherries are a fantastic natural source of melatonin, the hormone that helps regulate the sleep-wake cycle, and when included in your diet can have abundant effects on curing insomnia. Tart cherries also contain an essential amino acid, tryptophan, used by the body to create the B vitamin niacin, which plays a key role in producing serotonin, associated with mood and sleep.

# PITAYA ROSE

Often used as a romantic gesture sprinkled over your bed, dried rose petals add that same elegant touch to your moon milk.

## Ingredients

- 500 ml (16 fl oz) unsweetened almond milk
- 1 tablespoon raw cacao butter
- 1 tablespoon maple syrup
- ½ teaspoon ground cinnamon
- 1 teaspoon pink pitaya pea flower powder
- Dried rose petals (food-grade), to garnish

Any kind of dried rose petals can be used in this delicious moon milk recipe, with pink being the most popular. Dried rose petals can relieve stress and headaches, and also contain antioxidants, making this milk perfect to sip on at the end of a long day.

The pitaya used in the recipe, also known as dragonfruit, has recently stolen the superfood spotlight. Maybe one of the most unique-looking fruits, with its leather-like, pink and green skin and distinctive seed-speckled flesh, it is packed with antioxidants that help to reduce cholesterol, contain anti-inflammatory properties and stimulate the growth of new blood vessels.

Pink pitaya powder combined with dried rose petals is a perfect match made in moon milk heaven.

# MUSCLE MOON MILK

You've had a tense day, hit the gym hard to help relieve some stress, and now you're looking to replenish your muscles while calming your body, mind and spirit. This moon milk ticks all those boxes and sets you up for some serious zzz's.

## Ingredients

500 ml (16 fl oz) unsweetened oat milk
1 scoop unflavoured protein powder (see below)
1 tablespoon agave syrup
½ teaspoon ashwagandha powder

## Tip

This recipe uses unflavoured protein powder as a boost, but feel free to use your favourite plant-based protein powder. Vanilla is good, but try to avoid chocolate as it may contain a little caffeine, which counteracts sleep. Also be aware of your protein powder's sugar content and try to keep the sugar to a minimum.

When we sleep, our body does most of its regenerating and rebuilding, so set yourself up for an extra-restorative sleep with the use of your favourite protein powder mingled with ashwagandha powder, which further helps the body cope with stress and insomnia.

This is also a great way to get added protein if you're on a primarily plant-based diet. Protein is essential to our diet and supplies the building blocks for bones, muscles, cartilage, skin, blood, hair, hormones, enzymes and nails. Protein intake is certainly sustainable on a plant-based diet, from such things as tofu, soy, beans, legumes, grains, chia seeds, dark leafy greens, and potatoes. Yet while you may eat many of those food items daily, it's not a bad idea to add a protein supplement to your diet too – and what better way than in a delicious, comforting, muscle moon milk.

# INDEX

Recipes are in *italic*.

# SUPPLIERS

**UNITED KINGDOM**

PLANET ORGANIC
planetorganic.com

HOLLAND & BARRETT
hollandandbarrett.com
*Ships internationally*

WHOLE FOODS
wholefoodsmarket.co.uk

AMAZON
amazon.co.uk

**AUSTRALIA AND NEW ZEALAND**

HONEST TO GOODNESS
goodness.com.au

NATURAL HEALTH ORGANICS
naturalhealthorganics.com.au
*Ships internationally*

WHOLEFOODS
wholefoodshealth.co.nz

**UNITED STATES**

WHOLE FOODS
wholefoodsmarket.com

SUNCORE FOODS
suncorefoods.com
*Ships internationally*

KROGER
kroger.com

AMAZON
amazon.com
*Ships internationally*

NATURAL GROCERS
naturalgrocers.com

TRADER JOE'S
traderjoes.com

FRESH THYME
freshthyme.com

THRIVE MARKET
thrivemarket.com

SPROUTS FARMER'S MARKET
sprouts.com

**EUROPE**

**Germany**
REFORMHAUS
reformhaus.de

ALNATURA
alnatura.de

FÜLLHORN
fuellhorn-biomarkt.de

**France**
BIOCOOP
biocoop.fr

BIO C' BON
bio-c-bon.eu

NATURALIA
naturalia.fr

**Italy**
NATURASI
naturasi.it

**Spain**
MERCADONA
mercadona.es

**The Netherlands**
ALBERT HEIJN
ah.nl

# ACKNOWLEDGEMENTS

First I would like to say how incredibly grateful I am for this opportunity. On my first date with my husband, I told him one day I was going to have a book published. I didn't know what kind of book, or when it would happen, but nearly ten years later here we are. I want to thank Quarto Publishing for their hard work, guidance, love and support. As a first-time author, I had no idea what to expect. They have been patient, kind and understanding as we worked through this book together (while supporting my need to soak up all the new-born snuggles as I welcomed my baby girl into the world right in the middle). Thank you to my husband who watched the kids as I made sometimes four moon milk recipes in one day, and then helped me out by enjoying them with me after the kids went to bed. He also made a really good photography assistant behind the scenes and offered his support fully as I worked on this very fun project. Thank you to my two-year-old son for being so good while 'mummy worked', and playing 'pretend moon milk' as I whipped up mine. I want to say thanks to the rest of my family and friends for their love and support. Thank you to all my fellow food bloggers and photographers who have been kind enough to share with me their tips and tricks and methods for food photography – I have learnt a great deal from them – and for their ongoing support. And lastly (but certainly should be first), I give all thanks and glory to God. He is the reason why I set out on this path to a healthier lifestyle, and I certainly would not be publishing a book, this very book, or feeling as great as I do now if He hadn't shown me the way to a healthier lifestyle.

*The Caramel moon milk recipe was inspired by Pinch of Yum's 5 Minute Magic Vegan Caramel Sauce.*